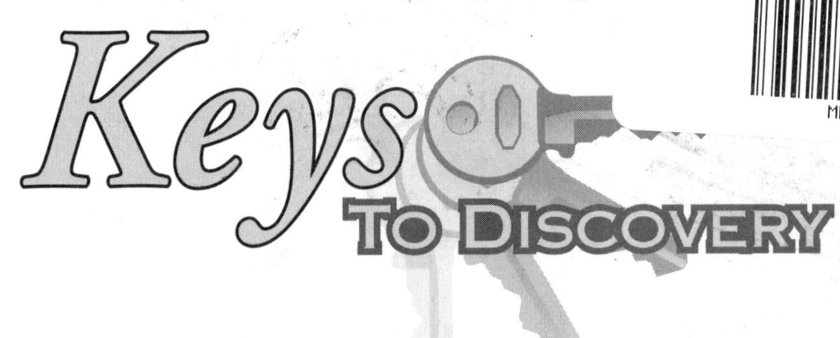


# Energy

## Contributing authors

Sara E. Freeman, Catherine Hernandez, Mel Feigen, Cynthia Nagel

## Contributing illustrators

Mark Mason, Eileen Mueller Neill, Marilynn Barr,
Marilee Harrald-Pilz, Catherine Yuh, and Anthony D. Paular

**This book is a compilation of these Frank Schaffer materials:**
FS-62019 Energy
FS-62020 Heat
FS-62022 Electricity
FS-62023 Light
FS-62024 Sound

FS-83130 Energy

23740 Hawthorne Blvd., Torrance, CA 90505
CAT. NO. FS-83130
ISBN 0-7682-0174-8

# Table of Contents

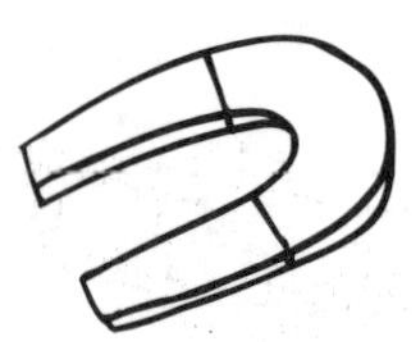

# Introduction

Welcome to *Energy,* a Frank Schaffer "Keys to Discovery" book designed to open the door to learning. Lively, upbeat activities balanced with generous, step-by-step guidelines bring even the most reluctant child across the threshold and onto the path of discovery.

Each section is organized around a list of concepts about energy and two "discovery through experiments" activities based on those concepts. The experiments, meant to be used at a center or with small groups, focus on such skills as observation, record-keeping, making predictions, critical thinking, and drawing conclusions. Children will find themselves using scientific methodology such as measurement, classification, exploring, using objects to test their theories, drawing what they see . . . and loving it! Three reproducible activity sheets in each of the five sections bring scientific discovery to the children's level of understanding.

Children will also have the opportunity to demonstrate how they can translate their newfound skills into other subject areas. Two pages of "curriculum connections" are included in each section, bridging science with activities in language arts, math, and art. From measuring changes in their shadows to making a papier mâché rattle, children will delight in the knowledge they gain as a result of their natural curiosity.

Loaded with reproducible teaching aids, *Energy* features an illustrated notebook page and an award certificate for the children, both specially designed to relate to the activities in the book and to bring on a smile. A letter to parents paves the way for family interaction on certain projects, sharing with these essential "teachers" the gist of activities and skills their children will be working on through the course of the book. A record sheet makes it easy for you to keep track of each child's progress. We hope you enjoy this product!

Dear Parents,

Over the course of the next few weeks, your child will be participating in a unit designed to build enthusiasm for science concepts. The class will be experiencing real-life science activities about energy, heat, electricity, light, and sound.

We will be working on developing science skills together. Every week we'll explore a science concept and do experiments, such as discovering how light travels. As your child progresses through the activities, he or she will become familiar with scientific methods such as measurement, classification, exploring, using objects to test their theories, drawing what they see, and much, much more.

I will be presenting a variety of science concepts and helping the children through their use with plenty of examples and experiments. Occasionally, I'll also send home a project for you to work on as a family. I greatly appreciate your interaction with your child on these projects. The certificate of award he or she brings home at completion of the unit is meant for you too!

Sincerely,

______________________________

(Teacher)

# Energy

## Concepts

Energy makes it possible for things to move and change. Energy can neither be created nor destroyed; but it is constantly changing. Children experience energy in many ways. They feel the sun's heat energy shining down on them. The food they eat gives them energy to work and play. They use electrical energy when working on a computer. Here are some beginning energy concepts for young students to learn:

- Energy can be stored (potential energy) or moving (kinetic energy).
- Energy can be transferred from one thing to another.
- There are many sources of energy. The sun is our most important one.
- Our bodies get energy from the food we eat.

## Discovery Through Experiments

### Energy—Stored and Moving

**Materials**
activity and record sheets (found on pages 8 and 9), a rectangular pan half-filled with wet sand, two tennis balls

**Exploration**
Each student will drop a ball from knee-height and another from above his or her head and then compare the dents the balls make when they land in the wet sand.

**Discovery**
The ball dropped from above the head makes a larger dent in the sand than the ball dropped from knee-height. The ball dropped from above the head had more energy. (You may wish to explain that the higher ball has more stored energy. As it falls, it picks up more speed than the lower ball. The higher ball is moving faster as it hits the sand.)

### Energy Transfer

**Materials**
activity and record sheets (found on pages 10 and 11), 12 to 18 dominoes

**Exploration**
Students will place 12 to 18 standing dominoes in a straight path. They will gently push the first domino and observe what happens. Students will then repeat the activity using a curved path of dominoes.

**Discovery**
When the first domino is pushed, it hits the second domino and transfers moving energy to it. Dominoes continue falling as they are hit by the dominoes before them and receive moving energy from those dominoes. The child's push transferred moving energy to the first domino.

 

# Language Arts Connection

Make a web to help organize the information your students learn about energy. Brainstorm words and ideas about energy and list them on the chalkboard. Then write the word *Energy* in large letters in the center of a piece of chart paper. Think of some categories and write them around the center. Have students help you find examples from the chalkboard for each category, and transfer them onto the chart paper.

**Energy**

**Kinds**
- chemical
- mechanical
- light
- heat
- sound
- electrical
- nuclear
- wind
- solar

- moving (kinetic)
- stored (potential)

**Machines**
- steam engine
- gasoline engine
- heater

**People**
- get energy from food
- use energy to live and to move

**Sources**

**Wind**
- sailboat
- windmill

**Water**
- waterwheel
- waterfall—electricity

**Sun**
- gives off light and heat
- makes plants grow
- makes rain and wind
- fossil fuels store its energy

**Saving Energy**
- turn off lights and TV when you leave a room
- walk, ride a bike, ride a bus, or carpool
- wear warmer clothes instead of turning up the heat

# Math Connection

## Food Chain Combinations

Have your students learn about food chains and math combinations with this activity. Direct pairs of students to draw and cut out a set of seven manipulatives—the sun, a wheat plant, a corn plant, a cow, a chicken, a pig, and a person.

Model for students a sample food chain: *The sun shines. A wheat plant uses energy from the sun to make sugars and starches. A cow eats the wheat and uses the wheat's stored energy. A person eats beef from the cow and uses the cow's stored energy.*

Have students use their manipulatives to make six different food chains, each involving the sun, a plant, an animal, and a person. Have students make an organized list to record their combinations.

- sun, wheat, cow, person
- sun, wheat, chicken, person
- sun, wheat, pig, person
- sun, corn, cow, person
- sun, corn, chicken, person
- sun, corn, pig, person

# Art Connection

## Wind Energy—Make a Pinwheel

**Materials**

white paper, pencil with eraser, scissors, crayons, thumbtack, metric ruler, tape

**Directions**

1. Cut out a white square with sides that are 15 centimeters. Fold it in half diagonally. Unfold. Fold diagonally in the other direction.

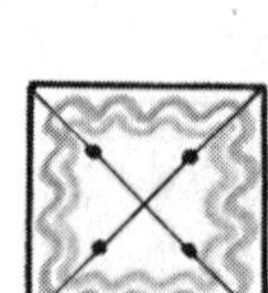

2. Color the front and back.

3. From each corner measure 7 centimeters in toward the center. Make a dot. Cut along the fold to the dot.
4. Bring half of a newly cut corner up and over the center point. Tape it. Repeat for the other three corners. Poke a thumbtack through the center into the pencil eraser.

5. Experiment to discover which way to blow to make your pinwheel turn best.

# Science Connection

## Make a Water Wheel

**Materials**
two-inch section of a cardboard paper tube, paper plate, scissors, tape, marker, water faucet

**Exploration**
Cut four rectangular pieces from the edge of the plate as shown. Fold them in half, smooth sides together, at the edge of the ribbed section. Make blades by taping the pieces to the tube as shown, spacing them evenly apart. Slide the marker through the tube. Hold the marker as you place the tube under the faucet and lightly run water on it.

**Discovery**
The water will fall on the blades and the tube will turn. A water wheel uses water power in the same way. The wheel is connected by belts or gears to the machine it runs.

## Solar Energy Collector

**Materials**
warm sunny day, three baby food jars with lids, black paint, cold water, thermometer (optional)

**Exploration**
Paint the outside of one jar black. Fill each jar with cold water. Place the black jar and another jar in a sunny spot outside. Place the third jar in the shade. After one hour, feel the temperature of the water in each jar (or measure it with a thermometer). Compare.

**Discovery**
The water in the black jar should be the warmest. The sun is a source of heat energy. Some homes use solar panels to collect energy from the sun to heat their water. The panels contain a black material because that color absorbs the most sunlight.

# Literature Connection

*The Science Book of Energy* by Neil Ardley (Gulliver, 1992)
Color photographs and step-by-step instructions highlight this collection of science experiments.

*The Sun: Our Nearest Star* by Franklyn M. Branley (T. Y. Crowell, 1988)
Simple text and friendly illustrations introduce the reader to the sun—our important source of light energy and heat energy.

*Energy* by Illa Podendorf (Childrens Press, 1982)
This New True Book features color photographs and clear text that tells about the different kinds of energy and ways people use energy.

*Why Doesn't the Sun Burn Out?* by Vicki Cobb (Lodestar, 1990)
Nine questions about energy are answered in this book. The comic-book-style illustrations show children exploring energy concepts.

*Arthur's New Power* by Russell Hoban (T. Y. Crowell, 1978)
In this humorous story, the power keeps going out at the Crocodile home. So Mother, Father, Arthur, and Emma Crocodile must all learn to conserve energy.

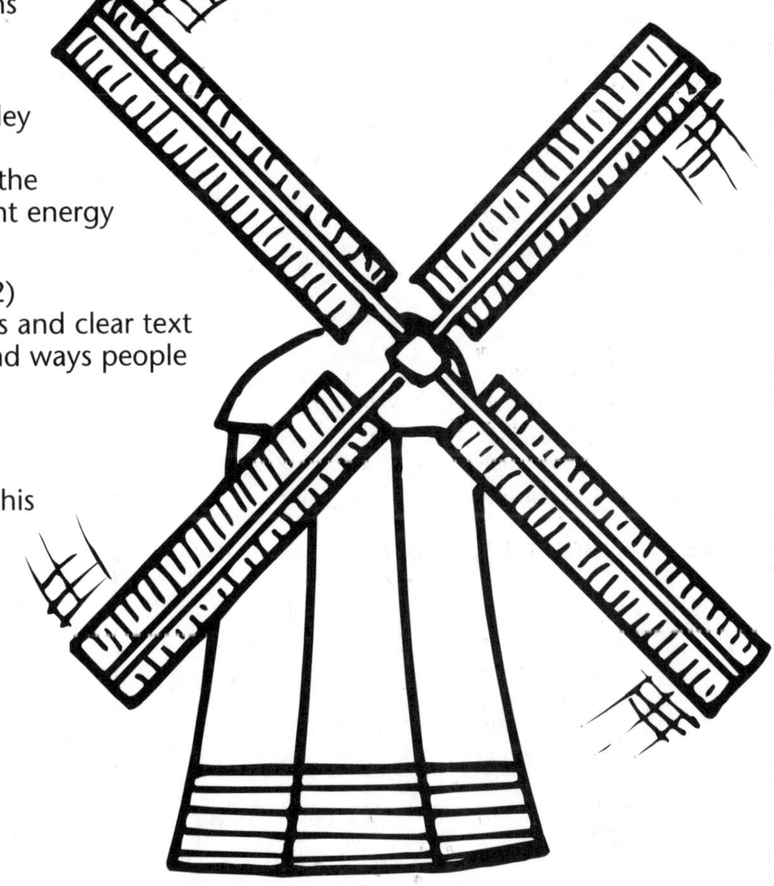

# Energy—Stored and Moving

**Question:** Does a ball dropped from knee-height have the same amount of energy as a ball dropped from above your head?

**Materials:** pan half-filled with wet sand
two tennis balls

**Predict:** On your record sheet, predict whether a ball dropped from a lower height will make a dent smaller than, equal to, or bigger than the dent made by a ball dropped from a greater height.

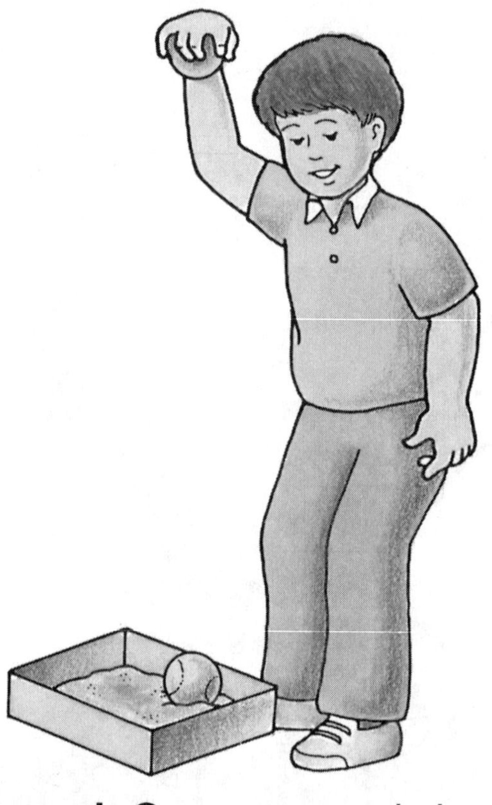

**Procedure:**

1. Stand by the pan. Hold one ball even with your knee and over the sand. Let it drop. Observe the dent it made in the sand.

2. Stand by the pan. Hold the other ball above your head and over the sand. Let it drop and land in a different part of the sand. Observe the dent it made.

**Record:** On your record sheet, draw pictures of what happened. Write about your results.

**Think and Write:** As you were holding each ball, it had hidden, or stored, energy. When you dropped each ball, the stored energy became moving energy. Which ball had more energy as it hit the sand? How can you tell that it had more energy? Write your answers on your record sheet.

Name ______________________________ Record Sheet

# Energy—Stored and Moving

**Question**: Does a ball dropped from knee-height have the same amount of energy as a ball dropped from above your head?

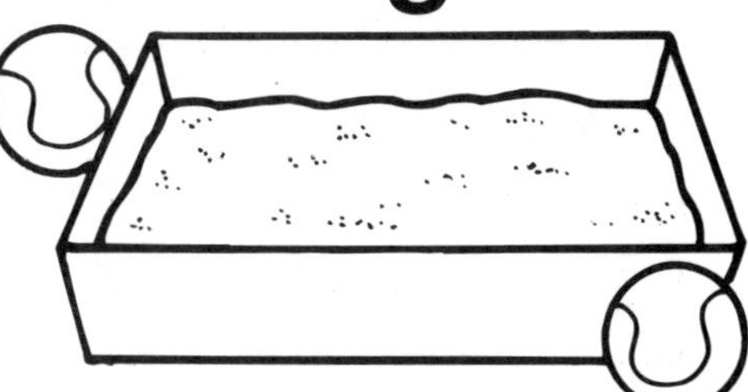

**Predict:** Will a ball dropped from a lower height make a dent smaller than, equal to, or larger than the dent made by a ball dropped from a greater height?

______________________________

______________________________

**Record:** Draw a picture of each dent. Write a sentence that describes it.

| 1. Dropped from knee-height | 2. Dropped from above the head |
|---|---|
| ______________________________ | ______________________________ |
| ______________________________ | ______________________________ |

**Think and Write:** As you were holding each ball, it had hidden, or stored, energy. When you dropped each ball, the stored energy became moving energy. Which ball had more energy as it hit the sand? How can you tell that it had more energy?

______________________________

______________________________

______________________________

 

# Energy Transfer

Activity Sheet

**Question:** When one standing domino is pushed into a line of standing dominoes, what will happen?

**Materials:** 12 to 18 dominoes

**Predict:** On your record sheet, predict what will happen.

**Procedure:**

1. Create a straight-line path of standing dominoes. Make sure each domino is no more than 1½ inches from the next domino.
2. Gently push the domino at one end of the path. Observe what happens.
3. Create a curved path of standing dominoes. This time, make sure that each domino is no more than 1 inch from the next domino.
4. Gently push a domino at the end of the path. Observe what happens.

**Record:** On your record sheet, write about what happened. Draw pictures of the straight-line path of dominoes.

**Think and Write:** When one thing moves, it can make a second thing move. The moving energy from the first thing transfers, or is passed on, to the second thing.
Where did the moving energy of the first domino transfer from?
Where did the moving energy of the second domino transfer from?
Where did the moving energy of the last domino transfer from?
Write your answers on your record sheet.

Name ____________________ Record Sheet

# Energy Transfer

**Question:** When one standing domino is pushed into a line of standing dominoes, what will happen?

**Predict:** What do you think will happen?

______________________________

**Record:** Write about what happened.

______________________________

______________________________

Draw the straight-line path of dominoes at the start, when your finger pushed the first domino.

Draw a picture at the end, after the last domino fell.

**Think and Write:** When one thing moves, it can make a second thing move. The moving energy from the first thing transfers, or is passed on, to the second thing.

Where did the moving energy of the first domino transfer from? ____________________

Where did the moving energy of the second domino transfer from? ____________________

Where did the moving energy of the last domino transfer from? ____________________

Name ______________________________ Homework

# Saving Energy

Electricity is a form of energy. People use electric lights, hair dryers, and refrigerators. Look around your home. Find and list eight things that use electricity.

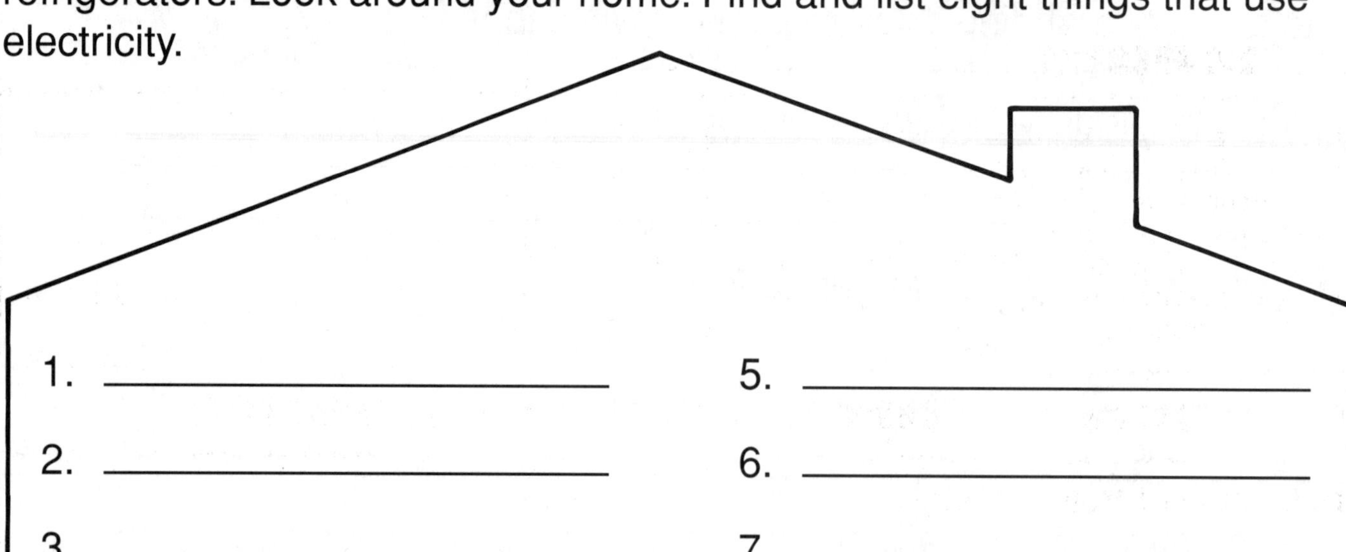

Electricity is made at power stations. The machines there often use coal or oil as fuel. Someday the earth may run out of coal and oil. So it is important to save, or conserve, energy. One way you can save energy is to turn off lights when you are not using them.

Color and cut out this switch plate. Tape it to a light switch at home.

# Heat

## Concepts

Heat is a form of energy children can easily explore. From feeling warm on a bright sunny day to watching an ice cube melt, children have everyday experiences that increase their understanding of heat and pique their interest to learn more. Here are some introductory concepts for young students:

- The sun heats the earth.
- Heat travels better through some objects than through others.
- Heat can change things.
- Friction produces heat.
- A thermometer measures the amount of heat an object contains.

## Discovery Through Experiments

### Heat Moves Through Objects

**Materials**
activity and record sheets (found on pages 16 and 17); two metal tablespoons; two wooden cooking spoons; warm, sunny spot; cup of ice cubes

**Exploration**
Students will place one metal spoon and one wooden spoon in a warm, sunny spot. After an hour they will touch the spoons to discover which one is warmer. In the meantime, students will place the other two spoons in the cup of ice cubes and touch them after two minutes to discover which spoon is colder.

**Discovery**
The metal spoon will get warmer in the warm spot as heat moves into it and cooler in the cool spot as heat moves out of it. Heat easily flows through metal, which is a good conductor. Heat does not easily flow through wood.

### Friction

**Materials**
activity and record sheets (found on pages 18 and 19), a piece of coarse sandpaper, a piece of wood

**Exploration**
Students will firmly rub their hands together for 30 seconds and then hold them to their cheeks to feel the heat they produced. Next students will briskly sand a piece of wood for one minute. They will then feel the sandpaper for changes in its temperature.

**Discovery**
Students will find that their hands are warmer after rubbing them together and the sandpaper is warmer after rubbing it against the wood. Friction is produced when one object rubs against another object. Friction is a source of heat.

 

# Language Arts Connection

On chart paper, make a word map about heat to organize information students have learned. Begin by writing the word *Heat* in the center in large letters. Choose four to six subheadings and write them around the heading. Then guide the children in thinking of examples for each subheading.

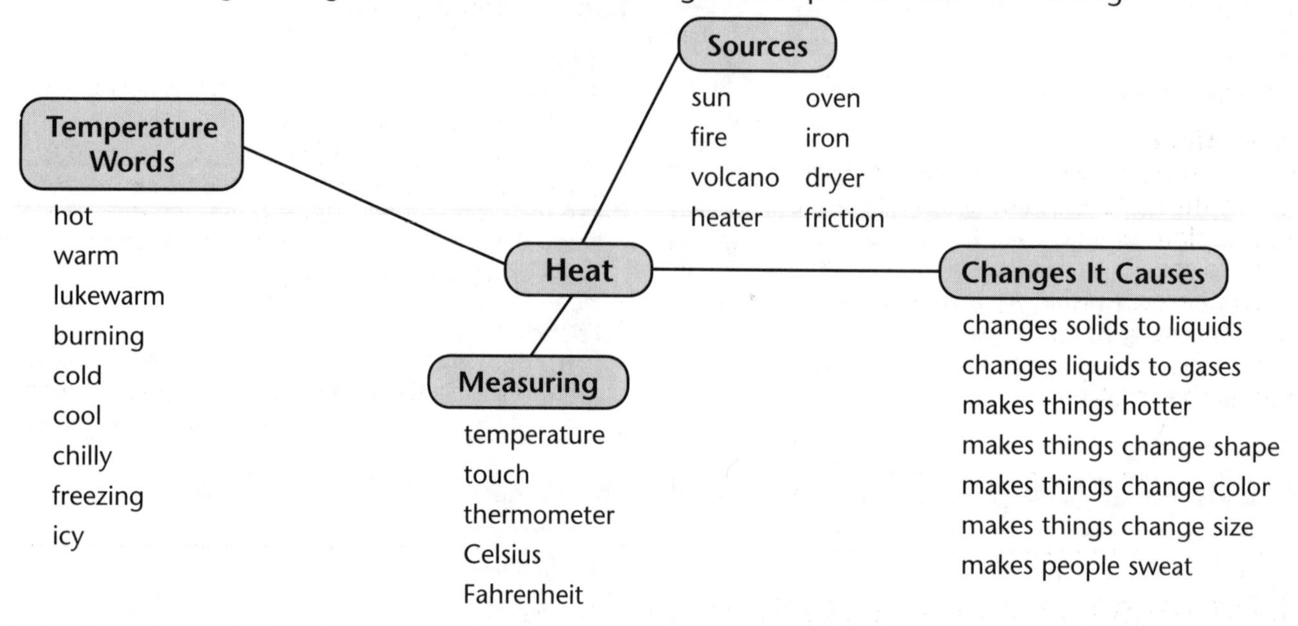


# Math Connection

## Reading a Thermometer

Have the children make their own thermometers using "Make Your Own Thermometer" on page 20. Then use the thermometer to practice these math concepts:

**Skip Counting**—The temperatures are shown in 10-degree increments. Begin at 0°F. Have students skip-count upward by tens. *(0°, 10°, 20° . . . )* Point out the smaller lines halfway between each number. Ask students what number is halfway between 0 and 10. *(5)* Have students begin at 0°F and skip-count upward by fives. *(0°, 5°, 10°, 15° . . . )*

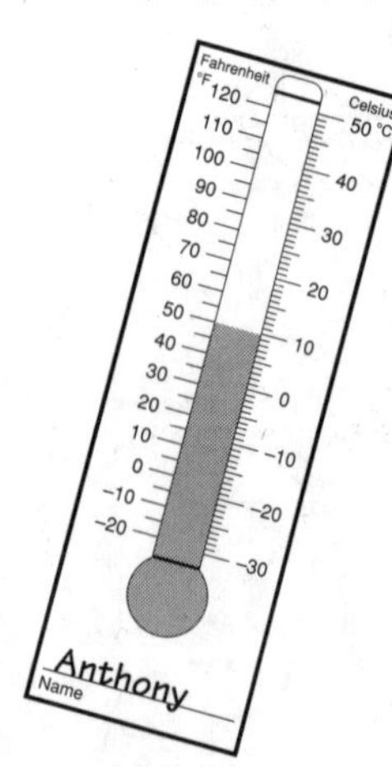


**Negative Numbers**—Teach the children that really cold temperatures go below zero and those numbers are shown with a negative sign. Begin at 50°C. Help the students count downward by tens to –30°. Ask them what temperatures would come next if the thermometer were to continue downward. *(–40°, –50°, –60° . . . )*

# Art Connection

## Colorful Campfire

This project uses heat to make a beautiful melted-crayon design that looks like a fire.

**Materials**

yellow, orange, red, and pink crayons; pencil sharpener; wax paper; scissors; brown construction paper; stapler; iron (for teacher use); thin rag

**Directions**

1. Use the pencil sharpener to make crayon shavings. Place the shavings inside a folded piece of wax paper.
2. (Teacher step) Put the rag on the wax paper and press with a warm iron until the crayons and paper melt together. When the wax paper has cooled give it back to the child.
3. Make the wax paper look like a fire by cutting it into a rough triangular shape.
4. Cut two strips of brown paper. Cross them to form the logs. Staple the logs to the base of the triangle to finish the campfire.

# Science Connection

## The Sun Is a Source of Heat

Take your class outside on a sunny day for this activity.

**Materials**
paper and pencil, thermometer (optional)

**Exploration**
In the morning, walk around the playground, having the children think about hot and cold as they explore different areas and surfaces. Direct them to record their observations of three different areas, surfaces, or objects. If you have a thermometer, measure the air temperature. Repeat the activity in the afternoon. Compare and discuss the results.

**Discovery**
Afternoon temperatures are usually warmer because the sun has been shining longer.

## Heat Changes Things

**Materials**
popcorn popper, popcorn

**Exploration**
Let each child examine an uncooked kernel of popcorn. Discuss different ways of making popcorn (frying pan and oil, microwave oven, electric popcorn popper) and the one element common to all—heat. Then make a batch of popcorn. Have the children compare their uncooked kernels to a cooked one.

**Discovery**
The kernel has a hard covering. When it is heated, moisture inside the kernel turns to steam. Pressure builds up inside until the kernel bursts, or "pops."

# Literature Connection

***The Science Book of Hot and Cold*** by Neil Ardley (Harcourt Brace Jovanovich, 1992)
Step-by-step directions and simple color photographs make this collection of experiments especially appealing.

***Burning and Melting*** by Peter Lafferty (Gloucester, 1990)
This nonfiction book uses clear text, color photographs, and simple diagrams to help explain concepts such as how heat is measured, how heat travels, and how materials expand when heated. Projects and quizzes keep the reader involved.

***The Popcorn Book*** by Tomie dePaola (Holiday, 1978)
This fun book gives science facts, recipes, and information on the history of popcorn, including different ways it was cooked by the Iroquois and other Native Americans.

***The Itch Book*** by Crescent Dragonwagon (Macmillan, 1990)
In this picture book, a hot, hot day in the Ozark Mountains gets everyone itching and scratching. Luckily, it's nothing that can't be solved by a night picnic at ice-cold King Creek.

***Chinook!*** by Michael O. Tunnell (Tambourine, 1993)
Two children new to town are captivated by an old-timer's tall tales of chinook winds so hot that in a flash they melt snow as high as rooftops, leaving dry parched ground.

 

# Heat Moves Through Objects

**Question:** Which will be warmer in a warm spot and cooler in a cool spot—a metal spoon or a wooden spoon?

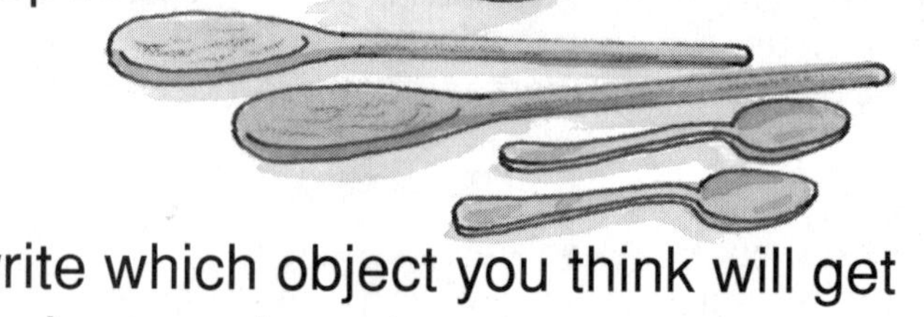

**Materials:** two metal tablespoons
two wooden cooking spoons
warm, sunny spot
cup of ice cubes

**Predict:** On your record sheet, write which object you think will get warmer in a warm spot and cooler in a cool spot.

**Procedure:**

1. Place one metal spoon and one wooden spoon in a warm, sunny spot. Leave them for one hour.

2. Place the other metal spoon and wooden spoon in the cup of ice cubes for about two minutes. Take them out. Which spoon feels colder?

3. After an hour, check the spoons in the sunny spot. Which spoon feels warmer?

**Record:** On your record sheet, write about what happened in each step.

**Think and Write:** Heat moves from warmer things to cooler things. Materials that let heat move through them easily are called conductors. Which kind of spoon—metal or wooden—lets heat move into it and out of it easily? Which kind of spoon doesn't let heat flow through it easily? Write your answers on the record sheet.

Name ______________________________ Record Sheet

# Heat Moves Through Objects

**Question:** Which will be warmer in a warm spot and cooler in a cool spot—a metal spoon or a wooden spoon?

**Predict:** Which object will get warmer in the warm spot and cooler in the cool spot? ______________________________

______________________________

**Record:**

| | |
|---|---|
|  1. Which spoon felt warmer after being in a warm, sunny spot for an hour? ______________________________ ______________________________ | 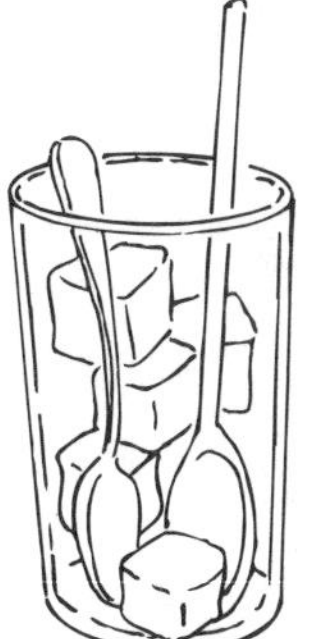 2. Which spoon felt cooler after touching the ice cubes for two minutes? ______________________________ ______________________________ |

**Think and Write:** Heat moves from warmer things to cooler things. Materials that let heat move through them easily are called conductors. Which kind of spoon—metal or wooden—lets heat move into it and out of it easily? Which kind of spoon doesn't let heat flow through it easily?

______________________________

______________________________

______________________________

# Friction

**Question:** When two objects rub against each other, will heat be produced?

**Materials:** your hands
coarse sandpaper
small piece of wood

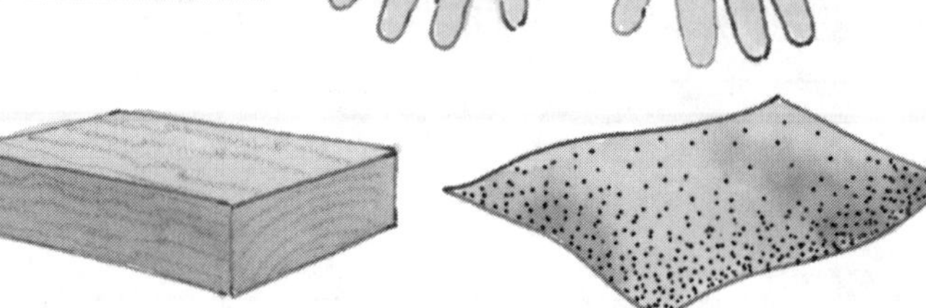

**Predict:** Will the objects listed produce heat when they are rubbed against each other? Write your prediction on your record sheet.

**Procedure:**

1. Put your hands against your cheeks. Do your hands feel cool, warm, or hot? Count to 30 as you rub your hands together as hard and as fast as you can. Put your hands against your cheeks. How do your hands feel now?

2. Touch the sandpaper. Does it feel cool, warm, or hot? Rub the sandpaper against the wood for one minute as hard and as fast as you can. Touch the sandpaper. How does it feel now?

**Record**: On your record sheet, describe the temperature of the objects before and after rubbing them.

**Think and Write:** Friction occurs when two objects rub against each other. What does this experiment teach you about friction? Write your thoughts on your record sheet.

Name ______________________________ Record Sheet

# Friction

**Question:** When two objects rub against each other, will heat be produced?

| | Your hands | Sandpaper and wood |
|---|---|---|
| **Predict:** How will the objects feel after being rubbed together? | | |
| **Record:** How do the objects feel *before* being rubbed together? | | |
| **Record:** How do the objects feel *after* being rubbed together? | | |

**Think and Write:** Friction occurs when two objects rub against each other. What does this experiment teach you about friction?

______________________________

______________________________

______________________________

______________________________

Name ______________________________

# Make Your Own Thermometer

1. Cut out the pieces. Color the circle at the bottom of the thermometer red.
2. Fold the thermometer in half down the center. Cut slits at lines **A** and **B**. Unfold.
3. Color one strip red. Leave one white. Glue them together at a tab to make one long strip.
4. Poke each end of the long strip through the thermometer slits. Then glue the ends together to make a loop.
5. Slide the red part of the strip up and down to show temperatures.

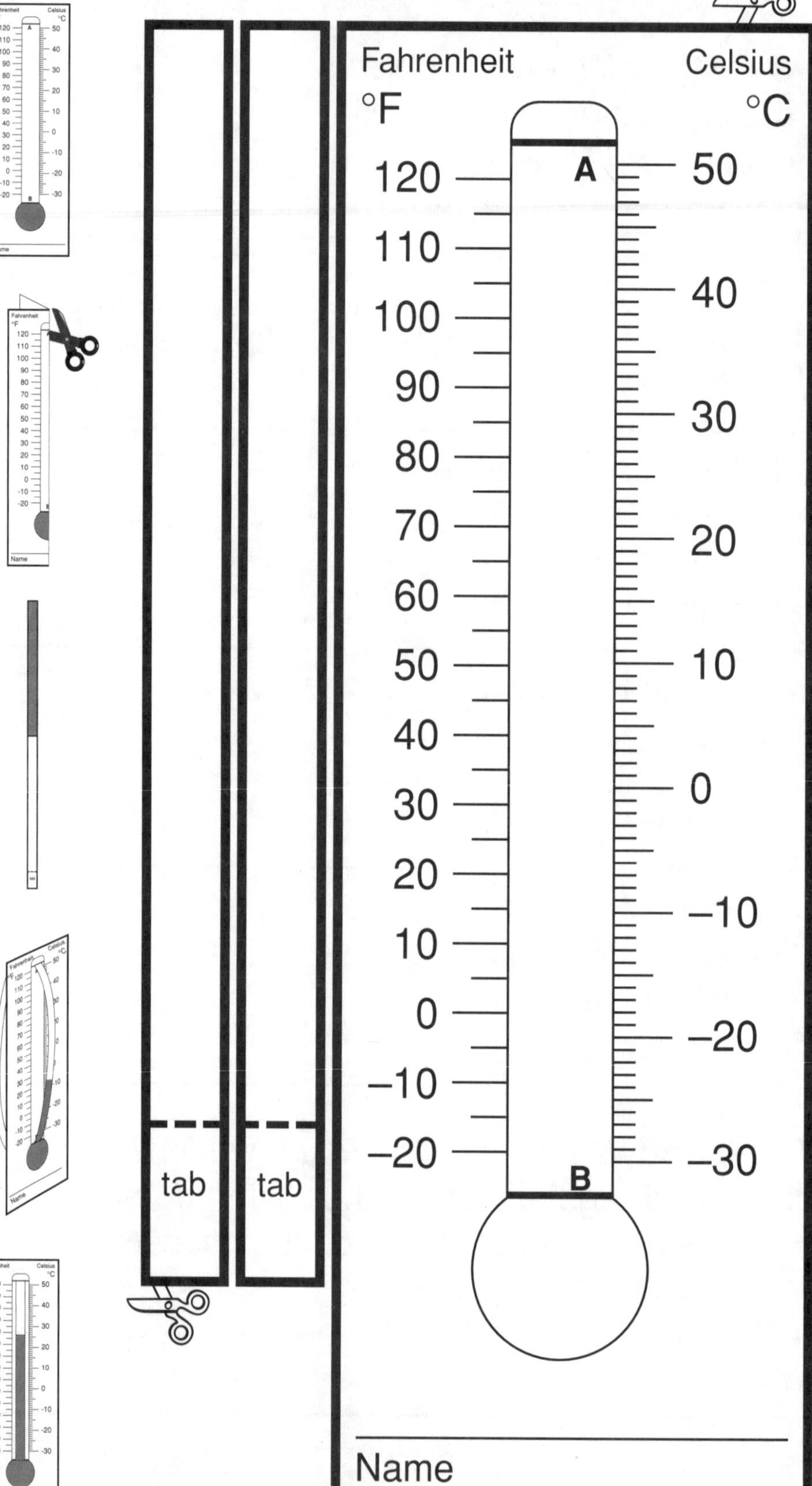

# Electricity

## Concepts

Electricity impacts the lives of children every day in countless ways—from the use of a simple lamp to light the book a child is reading to the use of a computer on which that same child may work and play. Explore the mystery of electricity with the activities in this unit. The following concepts are important to this topic:

- One kind of electricity (static electricity) is created when some materials rub together.
- Another kind of electricity flows into and out of things in a circle-like path (current electricity).
- Materials that allow electricity to flow through them are called conductors.
- It is very important that we follow safety rules when we use electricity. (See the reproducible activity on page 28.)

## Discovery Through Experiments

### Electricity That Jumps and Stops

**Materials**

activity and record sheets (found on pages 24 and 25), black paper, salt in a shaker, balloon, wool cloth

**Exploration**

Students will shake a small amount of salt onto the paper. Next they will stroke the balloon against the wool 25 times. Then they will hold the balloon over the salt, but not touching the salt.

**Discovery**

The balloon was electrically charged when it was rubbed against the wool. When the charged balloon was held over the salt, it attracted some of the salt, which stuck to the balloon. You may want to explain to your students that this kind of electricity is called *static electricity.*

### Electricity That Keeps Flowing

**Materials**

activity and record sheets (found on pages 26 and 27), a D battery (1.5 volts), a small flashlight bulb (2.5 volts), a 12-inch piece of insulated copper wire (called bell wire, available at hardware stores), masking tape

**Exploration**

Prepare the wire for students by baring its ends. To do this, cut around the insulation and then slide off the cut insulation. Students will twist one end of the wire around the bottom of the light bulb. They will then touch the other end to the battery and observe what happens. Next students will touch the bottom of the light bulb to one end of the battery and touch the loose end of the wire to the other end of the battery.

**Discovery**

The bulb will light only when it touches one end of the battery and the connecting wire touches the other end of the battery. The completed circuit allows electricity to flow.

 

# Language Arts Connection

## Using Electricity

Help your students categorize some of the specific ways electricity is used in daily life. Begin by writing the word *Electricity* and the headings shown below on a chalkboard or piece of chart paper. Then ask your students to think of items to list under each heading.

**Electricity**

| Makes Heat | Produces Motion or Power | Makes Light | Used in Communication |
|---|---|---|---|
| iron | washing machine | house lamps | television |
| toaster | cars, buses, trains | streetlights | telephone |
| stove | fan | flashlights | radio |
| blanket | clock | neon lights | motion pictures |
| incubator | escalator | | |
| | mixer | | |
| | vacuum cleaner | | |
| | drill | | |

# Math Connection

## Electrical Appliances Graph

Give students a homework assignment to list the kinds and number of electrical appliances in their homes. Then compile the data students collect onto a class graph.

**Directions**

On a large sheet of butcher paper, list the appliances your students found at their homes. Or glue magazine pictures of the various kinds of appliances to the graph. Then have each student paste onto the graph a small construction-paper square for each appliance that was in his home. For example, if there were seven lamps at home, then the child would paste seven squares next to the word *lamp*. When the graph is complete, ask your students these questions: *Which appliances are the most common in your homes? Which appliances are the least common? Are you surprised by what the graph shows?*

# Art Connection

## Nighttime Cityscape Mural

Involve your students in working together to make a mural that shows a city scene at night.

**Materials**

length of dark blue or purple mural paper, large sheets of construction paper in dark colors (black, brown, gray, olive green), tempera paints in bright colors

**Directions**

Show your students pictures of city scenes at night. Make sure students notice the dark buildings and backgrounds and the lit up windows and signs. Next have pairs or small groups of students make buildings of different sizes from rectangular sheets of construction paper. Show the children how to paint yellow or white windows on the buildings. Some students may want to paint lighted signs on the buildings. Then help your students glue the buildings to the mural paper. If desired, students can overlap some of the buildings for a more realistic look.

# Science Connection

## Conducting Electricity

Students will enjoy helping you test materials to determine which ones will conduct electricity.

**Materials**

D battery (1.5 volts), small flashlight bulb (2.5 volts), two 8-inch pieces of insulated copper wire (called bell wire, available at a hardware store), masking tape, small paper clips, toothpick, plastic spoon, strip of paper, aluminum foil, nail, rubber band, eraser, key

**Preparation**

Bare the ends of the wire pieces by using scissors to cut around the insulation, then sliding off the cut insulation. Twist one end of each wire piece around a small paper clip. Tape the other end of one piece of wire to the bottom of the battery. Twist the other end of the second piece of wire around the metal at the bottom of the light bulb. Lay the battery down and tape it to a secure surface. To test an item, tape one paper clip to one end of the item and the other paper clip to the other end. Then touch the bottom of the light bulb to the top of the battery. Explain to your students that if the bulb lights, it shows that electricity flowed through the item. If the bulb does not light, electricity did not flow through the item.

**Discovery**

Electricity will flow through the metal items. Tell your students that materials through which electricity flows are called conductors. Electricity will not flow through the rubber, wood, or plastic items. These items are insulators.

# Literature Connection

***The Science Book of Electricity*** by Neil Ardley (Gulliver, 1991)
Step-by-step directions for experiments help young readers understand concepts related to electricity.

***Discovering Electricity*** by Rae Bains (Troll, 1982)
Simple explanations about the importance, nature, and uses of electricity are included in this picture book.

***Switch On, Switch Off*** by Melvin Berger (HarperCollins, 1989)
How is electricity produced? How does it get into our homes? This title provides the answers to these questions in a style appropriate for primary-age children.

***Electricity*** by Graham Peacock (Thomson Learning, 1994)
This book presents numerous science activities that will help students discover how electricity works. Color photographs show children involved in each activity.

***Arthur's New Power*** by Russell Hoban (T. Y. Crowell, 1978)
When Arthur plugs in his amplifier, darkness envelopes the Crocodile family. In the days that follow, everyone in the family works at conserving electricity. Meanwhile, Arthur finds a solution of his own.

***How Many Stars in the Sky?*** by Lenny Hort (Tambourine, 1991)
A boy and his father escape the bright lights of the city so that they can count the stars in the night sky.

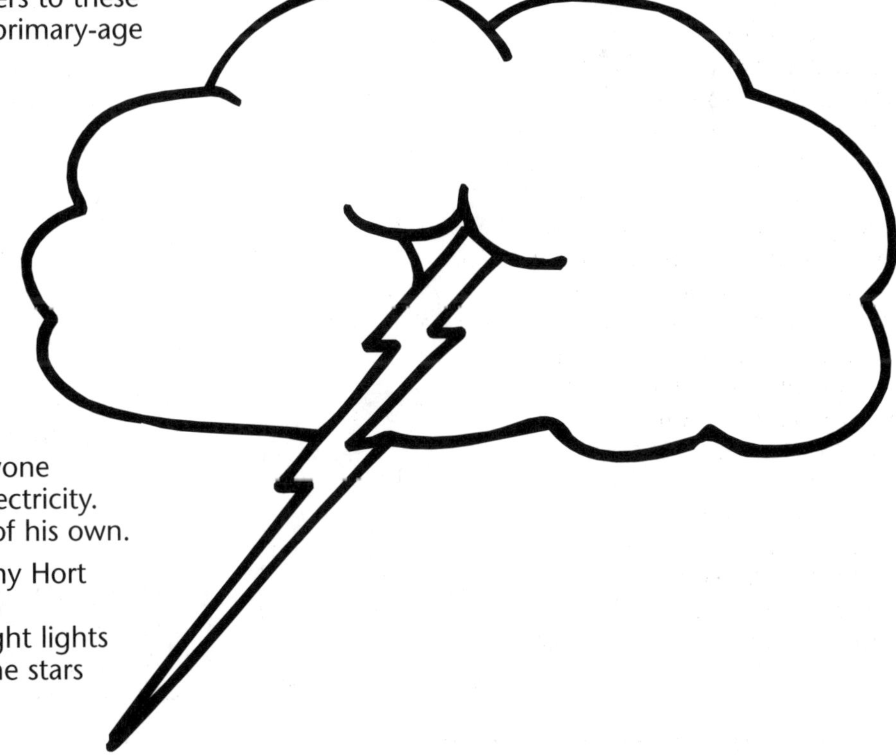

 

# Electricity That Jumps and Stops

**Question:** Can we make electricity by rubbing two things together?

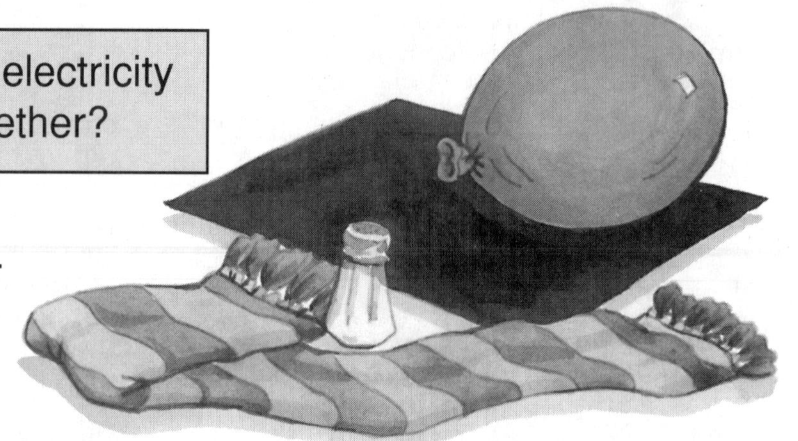

**Materials:** black paper
salt in a shaker
balloon
wool cloth

**Predict:** What will happen when a balloon that has been rubbed against a wool cloth is held over some grains of salt? Write your prediction on your record sheet.

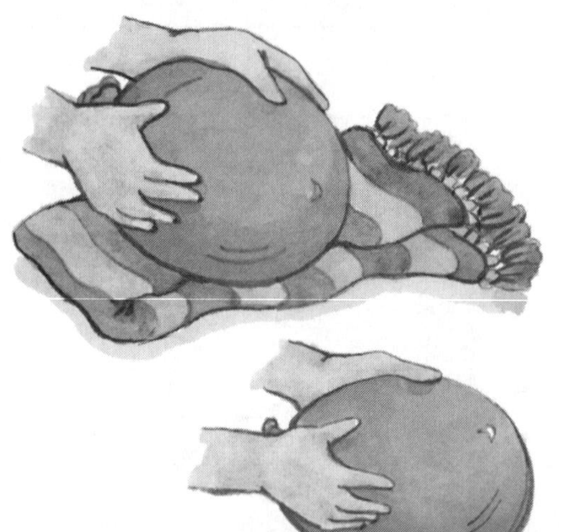

**Procedure:**

1. Shake a small amount of salt onto the black paper.
2. Stroke the balloon against the wool cloth 25 times.
3. Hold the balloon right over the salt. Do not touch the salt with the balloon.
4. Observe what happens to the salt. Turn the balloon over and look at it.

**Record:** On your record sheet, write about the results.

**Think and Write:** All things have tiny bits of electricity in them called electrons. When some things rub together, the electrons jump from one thing to the other.

Some electrons from the wool jumped to the balloon. This caused the balloon to be "charged" and attract the grains of salt. What do you think happens to some electrons in your hair when a comb rubs against it? Write your answer on your record sheet.

Name ________________________________________ Record Sheet

# Electricity That Jumps and Stops

**Question:** Can we make electricity by rubbing two things together?

**Predict:** What will happen when a balloon that has been rubbed against a wool cloth is held over some grains of salt?

______________________________________________

______________________________________________

**Record:** Write about what happened.

______________________________________________

______________________________________________

**Think and Write:** All things have tiny bits of electricity in them called electrons. When some things rub together, the electrons jump from one thing to the other.

Some electrons from the wool jumped to the balloon. This caused the balloon to be "charged" and attract the grains of salt. What do you think happens to some electrons in your hair when a comb rubs against it?

______________________________________________

Electrons are too tiny to be seen with your eyes. Draw a make-believe picture showing some electrons jumping from the wool to the balloon. Then draw the "charged" balloon over the salt.

| | |
|---|---|
| | |

# Electricity That Keeps Flowing

**Question:** How does electricity flow into a light bulb?

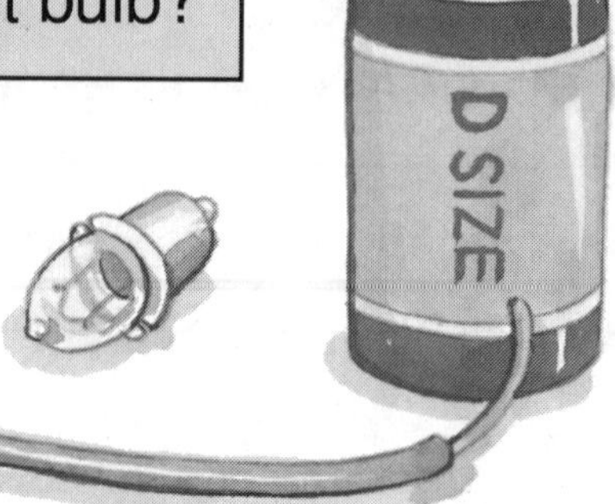


**Materials:** a D battery (1.5 volts)
a small flashlight bulb (2.5 volts)
a 12-inch piece of copper wire
masking tape

**Predict:** Will a bulb light up when wire connects it to one end of a battery? Or will a bulb light up when it touches one end of a battery and wire connects it to the other end of the battery? Write your prediction on your record sheet.

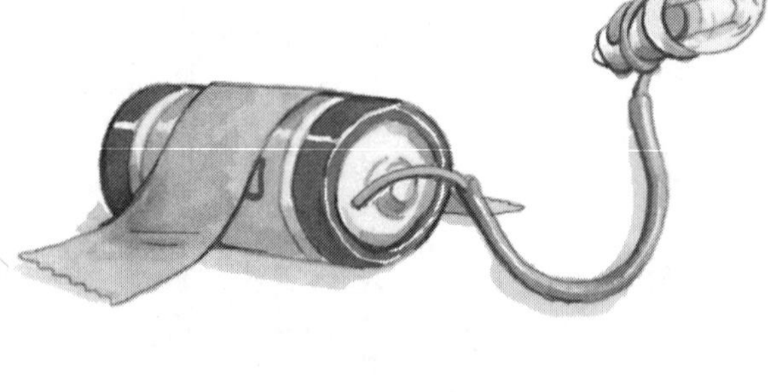

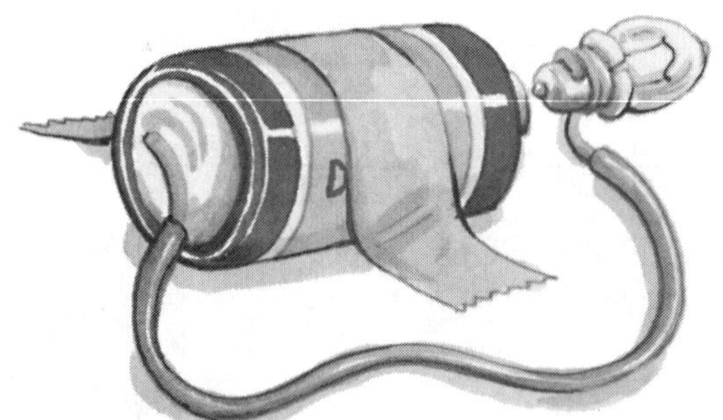

**Procedure:**

1. Lay the battery down and tape it to the table.
2. Twist one end of the wire around the metal bottom of the bulb.
3. Touch the loose end of the wire to one end of the battery. Observe the light.
4. Touch the bottom of the light bulb to one end of the battery. Touch the loose end of the wire to the other end of the battery. Observe the light.

**Record:** On your record sheet, draw pictures showing your results.

**Think and Write:** Electricity needs a circle-like path so it can flow into things and make them work. During which try did the bulb light up? Why do you think it lit up that time? Write your answers on your record sheet.

Name ______________________________ Record Sheet

# Electricity That Keeps Flowing

**Question:** How does electricity flow into a light bulb?

**Predict:** Will a bulb light up when wire connects it to one end of a battery?

Or will a bulb light up when it touches one end of a battery and wire connects it to the other end of the battery?

______________________________

______________________________

**Record:** Draw pictures showing your results.

| The bulb with wire connecting it to only one end of a battery | The bulb touching one end of a battery, with wire connecting it to the other end |
|---|---|
| | |

**Think and Write:** Electricity needs a circle-like path so it can flow into things and make them work. During which try did the bulb light up? Why do you think it lit up that time?

______________________________

______________________________

______________________________

______________________________

Name ______________________________ Cooperative learning

# Electrical Safety Rules

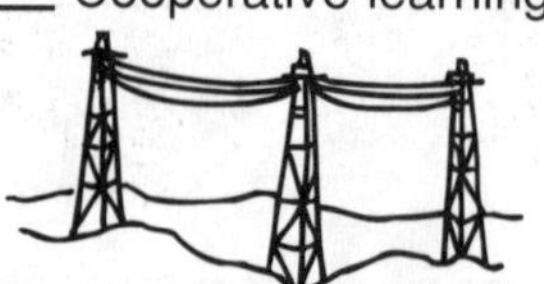

Work with a small group. Read the safety rules. Then choose one or two rules. Make up a skit that tells about each rule and why it is important. Plan your skit below.

**Safety Rules**

1. Never use the electricity from wall outlets for experiments.
2. Never play with wall outlets or wall switches.
3. Do not use an electric device in the bathtub, in the shower, or in a swimming pool.
4. Do not touch a switch or electric device when your hands are wet.
5. Stay away from power lines.
6. Do not fly a kite or model airplane near a power line.

Rules we choose ______________________________

______________________________

______________________________

Ideas for the skit ______________________________

______________________________

______________________________

______________________________

______________________________

Things we need for the skit ______________________________

______________________________

# Light

## Concepts

Light—including sunlight, lamplight, streetlights, firelight, and the dots of light that make a TV picture—is such an integral part of our everyday lives that we often take it for granted. Increase your students' understanding of light and its importance. The following concepts are included in this unit:

- Without light, we cannot see.
- Light travels in a straight line.
- Light readily passes through some materials.
- Only some light passes through other materials.
- Light is blocked by still other materials.
- White light is made up of colored light.
- A shadow is formed when an object blocks rays of light.

## Discovery Through Experiments

### The Way Light Travels

**Materials**
activity and record sheets (found on pages 32 and 33), four 4" x 6" index cards, flashlight, darkened area

**Exploration**
To prepare the center for your students, place a quarter in the middle of an index card. Trace around it and cut out the circle. Use the card as a template to make two more cards with holes in the same place. Fold in one inch on the sides of the four index cards. Students will line up the three cards with holes. They will shine the light through the hole in the first card and observe that it shines through all the holes. Then students will put the card without the hole in the place of the middle card. They will shine the light again through the first hole and observe the light.

**Discovery**
The light will be blocked by the middle card without the hole. Since light travels in straight lines, it cannot go around the card.

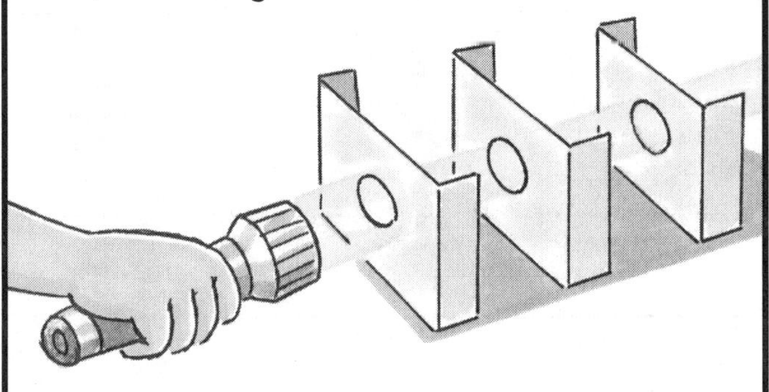

### Letting Light Through

**Materials**
activity and record sheets (found on pages 34 and 35), flashlight, clear drinking glass, copy paper, water in a clear cup, tissue paper, wax paper, plastic wrap, foil, cardboard, small piece of wood, darkened area

**Exploration**
Students will turn on the flashlight and shine it on each object. They will observe how much of the light passes through the object.

**Discovery**
Most of the light will pass through the glass, the plastic wrap, and the cup of water. Some of the light will pass through the copy paper, tissue paper, and wax paper. None of the light will pass through the foil, cardboard, and wood.

# Language Arts Connection

## "Light" Word Web

Stimulate your students to think of different words that describe light by having them complete this sentence: *The lights were* _______. Write the children's contributions on a word web like the one shown here. Keep the web on display in your classroom and encourage the children to refer to it for writing activities. For example, you may wish to have each child envision and write about a scene with lights, such as a city street, a city skyline, or a a fireworks display.

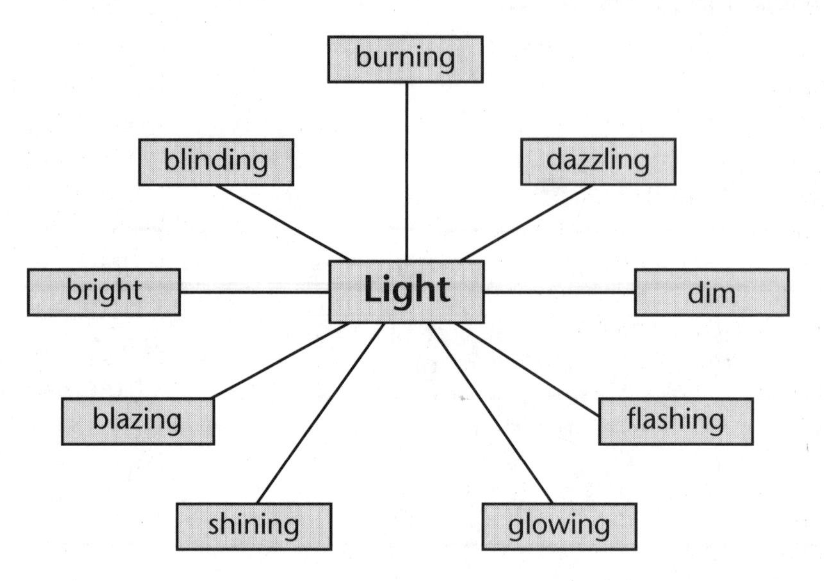


# Math Connection

## Measuring Shadows

Children will enjoy measuring and recording the changes in their shadows during the day. Plan to do this activity on a sunny day.

**Materials**
chalk, yardsticks and/or tape measures, pencils, paper

**Directions**
Give the children their own pencils and sheets of paper and take them outside in the morning. Have the children work with partners and give them these directions: Stand with your back to the sun while your partner uses chalk to mark the length of your shadow on the pavement. Then measure the length of your shadow and record the measurement and the time.

Repeat this activity two or three more times during the day. Near the end of the day, ask the students how the length of their shadows changed during the day. Students should find that their shadows were longer in the morning and afternoon, and shorter at midday.

# Art Connection

## Silhouettes

Use light and shadow to create individual silhouettes of your students.

**Materials**
filmstrip projector, white construction paper, black construction paper, pencil, scissors

**Directions**
Affix the white paper to a wall or bulletin board. Then have a child stand or sit in front of the paper so that his head is profiled. Shine the light from the filmstrip projector at the child, casting a shadow on the paper. Trace around the shadow, creating an outline. Then cut out the silhouettes or have your students cut them out. (If the children cut out the silhouettes, warn them to cut slowly and to be careful not to cut off braids, noses, and similar items.) When the silhouettes are cut out, have the children glue them onto black construction-paper backgrounds.

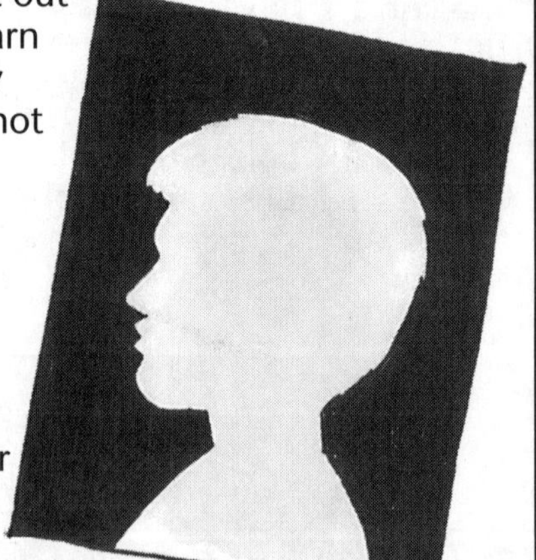

# Science Connection

## Sight and Light

Without light, we would not be able to see.

**Materials**
darkened classroom, colorful magazine pictures

**Exploration and Discovery**
Before darkening the classroom, ask the students to describe and note the color of some of the items in the classroom. Then darken the room and discuss with students how the colors in the room are no longer as discernable. (The degree to which items "lose" their color will depend on how much light is in your darkened room.) Pass around the magazine pictures and ask the children to comment on how well they can identify the colors in the pictures and what the pictures show. Turn on the lights and let the children observe the pictures again. Discuss with your students the fact that we need light to see.

## Looking at Shadows

Give your students an opportunity to explore the world of shadows.

**Materials**
white paper, flashlights for every two or three children (Some children may be able to bring flashlights from home. Or you may be able to borrow flashlights from other classrooms.)

**Exploration**
Partially darken the classroom and have the children shine flashlights on various small objects. Tell the children to try moving their flashlight closer to and farther away from the objects, and to also change the angle from which the light shines on the objects. Students may need to place a white sheet of paper under or behind the objects to see the shadows. Later, have a class discussion and allow the children to share any discoveries they made.

# Literature Connection

*The Science Book of Light* by Neil Ardley (Gulliver, 1991)
The step-by-step experiments featured in this book are sure to interest your young scientists.

*What Makes a Shadow?* by Clyde Robert Bulla (HarperCollins, 1994)
This book combines a text written to primary children with delightful illustrations that show children exploring shadows.

*Light and Darkness* by Franklyn M. Branley (T. Y. Crowell, 1975)
A discussion of basic properties of light, including how it enables us to see, is presented in this title.

*Fireflies in the Night* by Judy Hawes (HarperCollins, 1991)
A young girl and her grandparents enjoy the fireflies they see on a summer night. This gentle story is full of information about the beetles called fireflies.

*Light* by Graham Peacock (Thomson Learning, 1993)
This book answers a variety of questions about light and sight. Activities and experiments to help increase understanding of light and how it works are provided.

*Light* by Angela Webb (Franklin Watts, 1988)
Simple text and color photographs explore concepts about light for young children.

# The Way Light Travels

**Question:** Does light travel in straight lines or does it curve around things?

**Materials:** three 4" x 6" index cards with holes in the middle and the sides folded in
one 4" x 6" index card without a hole and with the sides folded in
flashlight
darkened area

**Predict:** Will light shining on index cards travel in a straight line or will it curve around the cards?

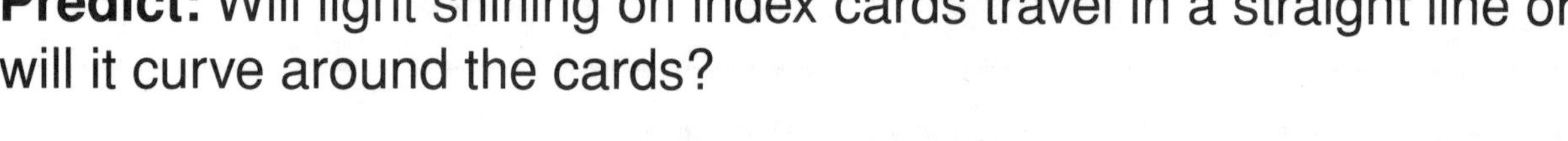

**Procedure:**

1. Line up the cards so that you can look through all three holes.
2. Shine the flashlight through the hole in the first card. Put your hand behind the third card. Does the light shine on your hand?
3. Take the middle card out. Put the card without a hole in its place.
4. Shine the flashlight through the hole in the first card. Put your hand behind the third card. Does the light shine on your hand?

**Record:** On your record sheet, draw pictures showing what happened to the light each time you shined it through the first hole.

**Think and Write:** Did the light curve around the card without a hole to shine through the next hole? Does light travel in straight lines or does it curve around things? Write your answers on the record sheet.

Name ______________________________ Record Sheet

# The Way Light Travels

**Question:** Does light travel in straight lines or does it curve around things?

**Predict:** Will light shining on index cards travel in a straight line or will it curve around the cards?

______________________________

______________________________

**Record:** Draw pictures showing what happened to the light each time you shined it through the first hole.

The light shining on the first set of cards

The light shining on the second set of cards

**Think and Write:** Did the light curve around the card without a hole to shine through the next hole? Does light travel in straight lines or does it curve around things?

______________________________

______________________________

______________________________

 

# Letting Light Through

**Question:** How much light passes through different objects?

**Materials:** flashlight, plastic wrap, copy paper, foil, tissue paper, darkened area, wax paper, clear drinking glass, wood, cardboard, water in a clear cup

**Predict:** How much light will pass through each object listed on your record sheet? Write your predictions on the record sheet.

**Procedure:**

1. Work in a darkened area. Turn on the flashlight and shine it on each object.
2. Observe each object as the light shines on it.

   Does most of the light shine through the object?

   Does only some of the light shine through it?

   Does none of the light shine through it?

**Record:** On your record sheet, write about the results.

**Think and Write:** What is one object that is useful to people because it lets a lot of light shine through? What is one object that is useful to people because it does not let any light shine through? Write your answers on the record sheet.

Name ____________________ 

# Letting Light Through

**Question:** How much light passes through different objects?

| Object | **Predict:** How much light will pass through the object?<br>• almost all of the light<br>• only some of the light<br>• none of the light | **Record:** How much light passed through the object?<br>• almost all of the light<br>• only some of the light<br>• none of the light |
|---|---|---|
| wax paper | | |
| glass | | |
| plastic wrap | | |
| copy paper | | |
| foil | | |
| water | | |
| cardboard | | |
| tissue paper | | |
| wood | | |

**Think and Write:** What is one object that is useful to people because it lets a lot of light shine through? What is one object that is useful to people because it does not let any light shine through?

_______________________________________________

_______________________________________________

_______________________________________________

Name ______________________________ Personal response

# Light Up Your Life

Complete each sentence. Color the light bulb and cut it out.

**Teacher:** Have the children color and cut out the light bulbs. Display them on a bulletin board titled "Lighting Up Our Lives."

# Sound

## Concepts

Sounds are a constant source of information about our world. They tell us what is happening in our environments. The sounds of a purring cat or a lullaby are pleasant; the sounds of a barking dog or a fire engine siren may seem harsh. The activities in this unit will introduce your students to the following basic concepts about sound:

- Sounds are caused by vibrations.
- Sounds travel through solids, liquids, and gases.
- Sounds travel in waves.
- Sounds range from soft to loud (intensity) and from low to high (pitch).

## Discovery Through Experiments

### Vibrations and Sound

**Materials**
activity and record sheets (found on pages 40 and 41), 12-inch plastic ruler, table or desk

**Exploration**
Students will hold the ruler at the edge of the table so that most of the ruler is off the table. Then they will hit the ruler, observe it vibrate, and listen to the sound it makes. Students will repeat the procedure with only four inches of the ruler off the table.

**Discovery**
The sound made when the ruler is hit the first time is lower than when the ruler is hit the second time. The ruler vibrates faster the second time, and faster vibrations result in higher-pitched sounds.

### Sound Travels Through Things

**Materials**
activity and record sheets (found on pages 42 and 43), a partner, two metal spoons, a large pot of water, a 30-inch length of yarn with finger-size loops tied at each end

**Directions**
First, one child will tap two spoons together while his partner listens from the other side of the room. Then the child will put his ear next to a pot of water while the other child puts the spoons underwater and taps them together. Finally, the child will tie the spoon to the middle of the yarn, put his fingers through the loops, and place his fingers in his ears. Then he will bend forward and his partner will tap the spoons together.

**Discovery**
The sound of the spoons tapping travels through air, water, and the piece of yarn. Children will enjoy the melodious chime-like sound made when the spoon tied to the yarn is tapped.

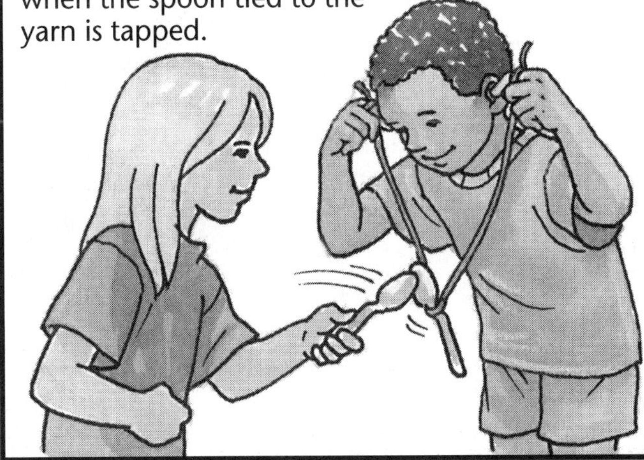

# Language Arts Connection

## Sounds Word Web

This activity will help increase your students' awareness of the great variety of sounds that fill their lives. Begin by writing the word *Sound* on chart paper and writing around it adjectives that describe sounds. Then ask your students to name things that make each kind of sound, and add them to the web.

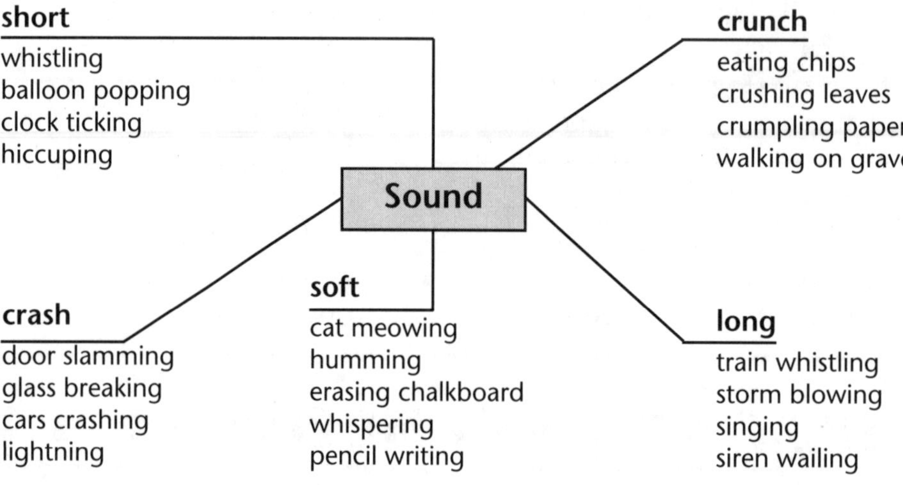


# Math Connection

## Auditory Mental Math

Engage your students in mental math practice with this auditory activity.

**Directions**

Collect a few glass jars. Fill one with one-quarter cup of water. Fill another with three-quarters cup of water. Add different colors of food coloring to each jar. Then assign a numerical value to each jar. For example, the jar with more water would have a value of five and the jar with less water would have a value of one. Tap the jars a few times with a spoon and have your students mentally add the numerical values indicated by each time you tapped a jar. For example, three taps on the "five" jar and one tap on the "one" jar would equal 16 (5 + 5 + 5 + 1). Then let your students take turns tapping the jars. Depending on the level of your class, you may want to add more jars with other numeric values.

# Art Connection

## Papier-mâché Shakers

Your students will enjoy making their own rattles. Then play some lively music and let your students shake away!

**Materials**

small cardboard tube; dried pasta, beans, rice, or corn; sheet of newsprint; masking tape; colored tissue paper; papier-mâché paste (one part flour to two parts water)

**Directions for students**

1. Cut a small piece of newsprint and tape it over one end of the tube.
2. Put a spoonful of pasta, beans, rice, or corn in the tube. Seal the other end with newsprint and tape.
3. Make one papier-mâché layer using torn strips of newsprint (1" x 4"). Hint: Run newsprint through fingers to remove excess paste.
4. Make another papier-mâché layer by carefully positioning torn tissue pieces (1" x $1\frac{1}{2}$") and then "painting" them on with a small amount of paste.
5. Allow several days for the shakers to dry.

# Science Connection

## Sounds All Around

Introduce your class study of sound with this auditory activity.

**Materials**
tape recorder

**Exploration and Discovery**
Make a tape recording of household sounds such as a doorbell, a telephone, a washing machine, a faucet, and a television. Leave a few seconds of blank tape after each sound. Play the tape for your students and have them identify the sounds. You may also want to make tapes of classroom and playground sounds. Explain to your students that a sound is made by something vibrating. Help them identify what may be vibrating and causing each sound.

## Watching Waves

**Materials**
plastic tub, water, four marbles wrapped together in plastic wrap

**Exploration**
Have the students fill the tub with two inches of water. Then give each child an opportunity to drop the plastic-wrapped marbles into the center of the water.

**Discovery**
Each child will see ripples, or waves, that travel outward from where the marbles entered the water. Sound waves travel outward from a source of sound in a similar manner. They travel in all directions.

# Literature Connection

***Sound*** by Graham Peacock (Thomson, 1993)
This book is filled with concepts about sound and suggestions for engaging children in their own explorations of the concepts. The illustrations show photographs of children trying out the activities.

***Sound and Light*** by David Glover (Kingfisher, 1993)
Experiments and explanations of the properties of sound and light are presented in this book.

***Train Song*** (T. Y. Crowell, 1990) and ***Plane Song*** (HarperCollins, 1993) by Diane Siebert
Lyrical, rhyming text evokes the movement and sounds of trains and planes.

***Who Says a Dog Goes Bow-wow?*** by Hank De Zutter (Doubleday, 1993)
The sounds of animals, as represented in languages all over the world, are featured in this book.

***Sing to the Stars*** by Mary Brigid Barrett (Little, Brown, 1994)
A boy with a gift for playing the violin teams with his elderly neighbor for a touching summer concert.

***Buffy's Orange Leash*** by Stephen Golder and Lise Memling (Kendall Green, 1988)
This story tells about how a hearing-ear dog helps his deaf owners.

# Vibrations and Sound

**Question:** Does the length of an object affect the sound it makes when it vibrates?

**Materials:** 12-inch plastic ruler
table or desk

**Predict:** What kind of sound will be made when a ruler hanging over a table edge is struck? Will the sound change when the amount of the ruler hanging over the table edge is changed?

**Procedure:**

1. Place a ruler on the edge of a table so that nine inches of it is off the table. Hold the ruler on the table with one hand.

2. Firmly strike the end of the ruler with two fingers. How does the ruler look? How does it sound?

3. Move the ruler so that only four inches of it is off the table.

4. Strike the ruler again. How does the ruler look? How does it sound?

**Record:** On your record sheet, write about your results.

**Think and Write:** Sounds are made when things vibrate. The second time the ruler was struck it vibrated faster than the first time it was struck. How did the faster vibrations change the sound? Write your answer on your record sheet.

Name ________________________________ Record Sheet

# Vibrations and Sound

**Question:** Does the length of an object affect the sound it makes when it vibrates?

**Predict:** What kind of sound will be made when a ruler hanging over a table edge is struck? Will the sound change when the amount of the ruler hanging over the table edge is changed?

______________________________________________

______________________________________________

**Record:** Write about your results.

| | | |
|---|---|---|
| The ruler with **nine** inches hanging over the table edge | What I saw | What I heard |
| The ruler with **four** inches hanging over the table edge | What I saw | What I heard |

**Think and Write:** Sounds are made when things vibrate. The second time the ruler was struck it vibrated faster than the first time it was struck. How did the faster vibrations change the sound?

______________________________________________

______________________________________________

______________________________________________

 

# Sound Travels Through Things

**Question:** Does sound travel through air? Through water? Through a piece of yarn?

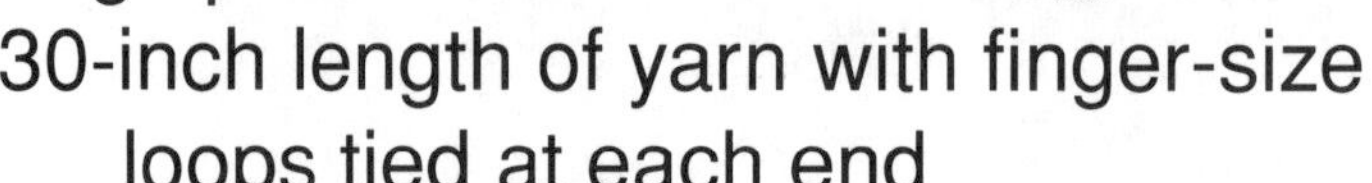

**Materials:** a partner
two metal spoons
large pot of water
30-inch length of yarn with finger-size loops tied at each end

**Predict:** Will the sound of two spoons clanging travel through air? Through water? Through a piece of yarn? Write your predictions on your record sheet.

**Procedure:**
Work with a partner. After each step, switch places with your partner and repeat the step.

1. Have your partner stand on the other side of the room and tap two spoons together.
2. Put your ear against the side of a pot of water. Have your partner put the spoons underwater and tap them together.
3. Tie a spoon to the middle of the yarn. Put a finger through each loop and then in each ear. Lean forward so the spoon hangs freely. Have your partner tap the spoon.

**Record:** On your record sheet, write about the results.

**Think and Write:** Sound travels through solids, liquids, and gases. Which time was the sound of the clanging spoons the loudest? Why do you think it was the loudest? Write your answers on your record sheet.

Name ______________________________ 

# Sound Travels Through Things

**Question:** Does sound travel through air? Through water? Through a piece of yarn?

**Predict:** Will the sound of two spoons clanging travel through air? Through water? Through a piece of yarn?

______________________________

______________________________

**Record:** Describe what you heard during each step of the experiment.

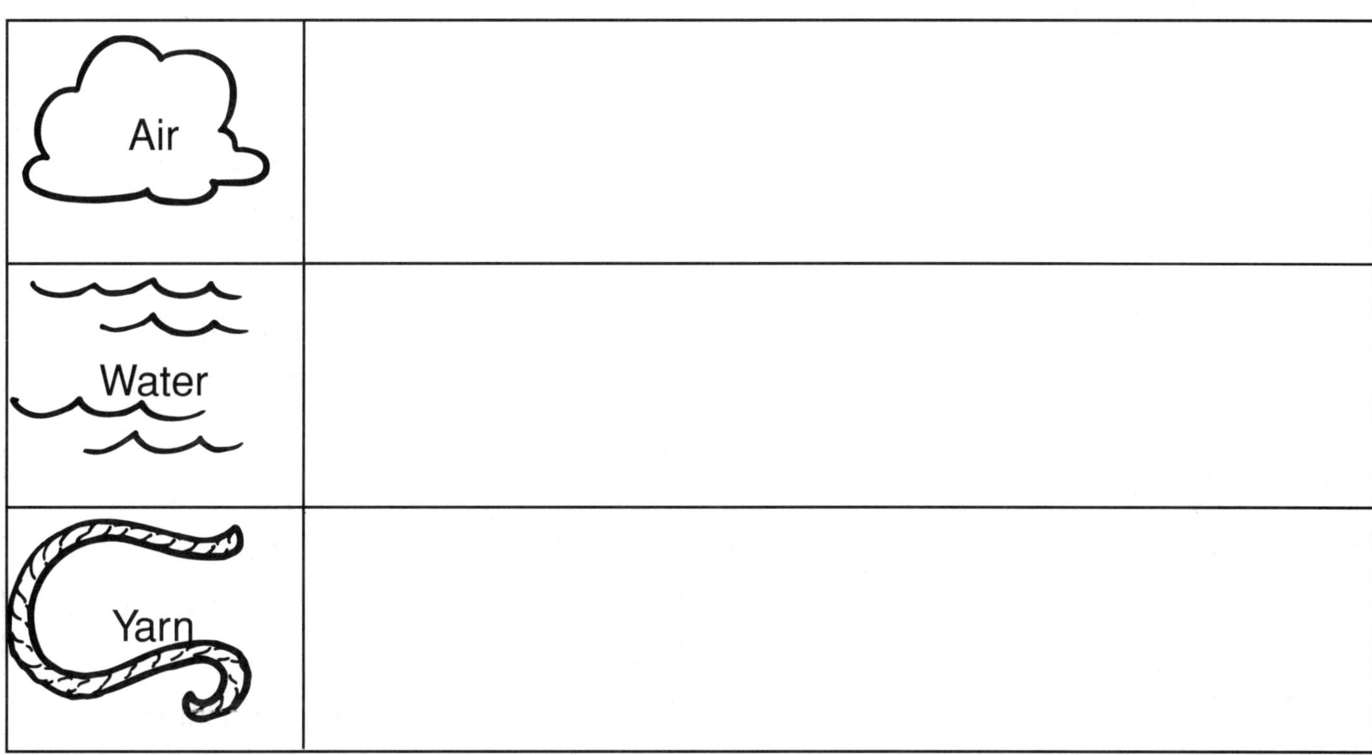

| | |
|---|---|
| Air | |
| Water | |
| Yarn | |

**Think and Write:** Sound travels through solids, liquids, and gases. Which time was the sound of the clanging spoons the loudest? Why do you think it was the loudest?

______________________________

______________________________

______________________________

Name ______________________________ 

# Listen and Record

Choose a place to sit and listen to sounds for five minutes.

Where did you sit? ______________________________

What time did you start listening? ______________________

In the boxes below, list things that you heard.

| Loud, high sounds | Soft, high sounds |
| --- | --- |
| | |
| **Loud, low sounds** | **Soft, low sounds** |
| | |

Which box has the most things listed in it? ______________________

Which sound did you like best? ______________________

# Congratulations!

You have discovered
the Keys to Success!

Keys TO DISCOVERY

Date

Teacher

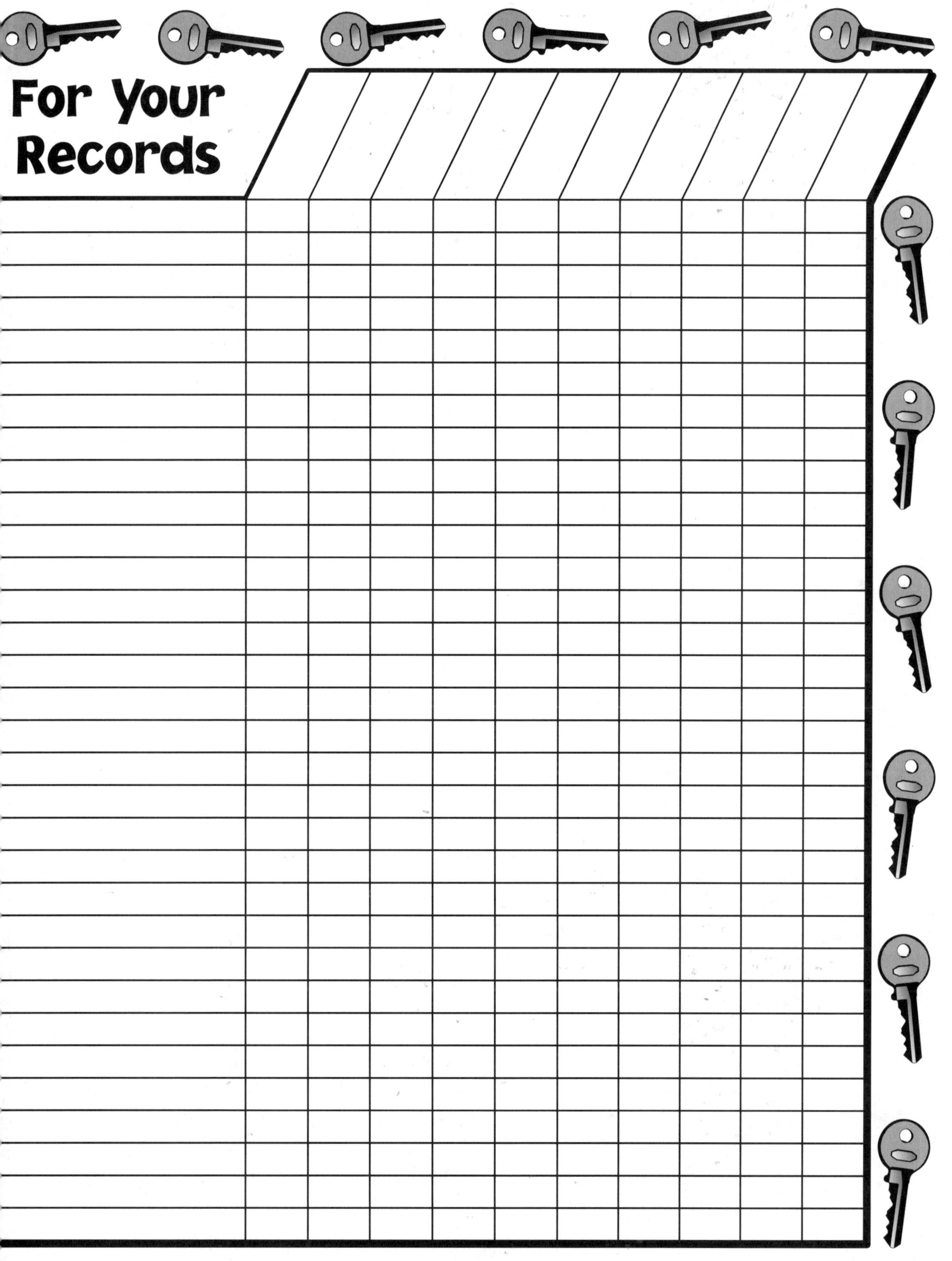
For Your Records

# THE SCIENTIFIC METHOD
## Keys TO DISCOVERY

### 1 Choose a problem.

What happens to a person's heart rate when he or she plays video games?

### 2 Research your problem

The heart beats faster when someone is mad, scared, or excited.

### 3 Develop a hypothesis.

If a person plays video games, then his or her heart rate will increase.

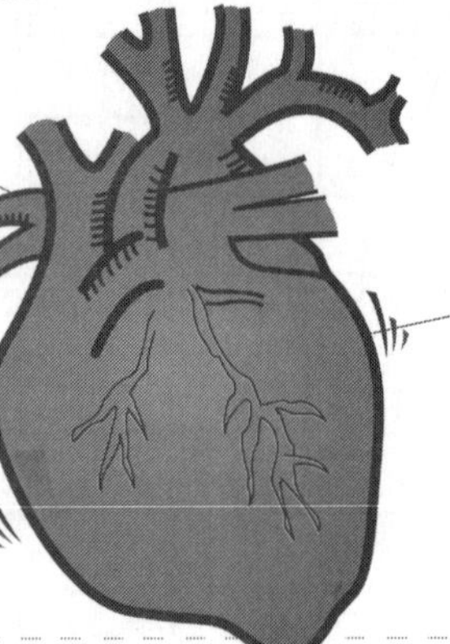

### Write your procedures. 4

a. Measure heart rate before playing a video game.

b. Measure heart rate while playing.

c. Measure heart rate after playing.

### 5 Test your hypothesis.

a. heart rate before playing – 72 beats per minute

b. heart rate while playing – 96 beats per minute

c. heart rate after playing – 76 beats per minute

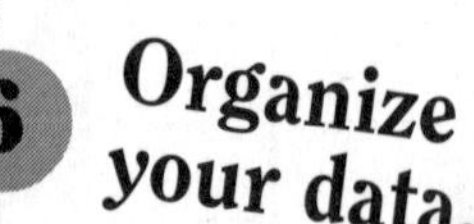


### 6 Organize your data.

**TABLE OF FACTS**

| | Heart rate before playing | Heart rate while playing | Heart rate after playing |
|---|---|---|---|
| Student 1 | 72 | 96 | 76 |
| Student 2 | 70 | 89 | 74 |
| Student 3 | 74 | 100 | 80 |
| Student 4 | 72 | 80 | 74 |

### 7 State your conclusions.

Playing video games increases a person's heart rate.